W9-BTH-091

CAMOUFLAGE

CAMOUFLAGE

NATURE'S DEFENSE

NANCY WARREN FERRELL

A FIRST BOOK 1989

FRANKLIN WATTS NEW YORK
LONDON TORONTO SYDNEY

TO CAROLYN AND AMOS

Special thanks to
Mary Lou King; to James King, Biologist (retired),
U.S. Fish and Wildlife; and to Philip Schempf,
Project Leader, U.S. Fish and Wildlife, Juneau, Alaska

Cover photograph courtesy of
Animals/Animals (Harry Engels)
Illustrations by Anne Canevari Green

Photographs courtesy of:
National Marine Fisheries Service: pp. 3 and 19
(R. Carlson); Alaska Fish & Game: pp. 8, 15 (both,
Bob Armstrong), 17 (both, Bob Armstrong), 24 and
25 (J. Hyde), 51 (Bob Armstrong); Animals/Animals:
pp. 11 (Ted Levin), 28 (Zig Leszczynski), 31, 35 (Zig
Leszczynski), 38 (Marcia W. Griffen), 40 (top, Oxford
Scientific Films, Michael Fogden; bottom, Zig
Leszczynski), 44 (top, E.R. Degginger; bottom,
George K. Bryce), 46 (Zig Leszczynski), 49
(E.R. Degginger), 52 (Zig Leszczynski).

Library of Congress Cataloging-in-Publication Data

Ferrell, Nancy Warren.
Camouflage/Nancy Warren Ferrell.
p. cm.—(A First book)
Bibliography: p.
Includes index.
Summary: Discusses various forms of protective defenses used by
animals, with an emphasis on camouflage phenomena such as disruptive
coloration and countershading.
ISBN 0-531-10688-8
1. Camouflage (Biology)—Juvenile literature. 2. Animal defenses—
Juvenile literature. [1. Camouflage (Biology) 2. Animal defenses.]
I. Title. II. Series.
QL767.F43 1989
591.57'2—dc19 88-38064 CIP AC

Printed in the United States of America
6 5 4 3 2 1

CONTENTS

CAMOUFLAGE

This marmot—a kind of woodchuck—blends well with the rocky hillsides where it lives.

THE HIDE-AND-SEEK WORLD

People play hide-and-seek as a game, but animals do not. To the animals, it means their very survival.

On the one hand, nature helps animals stay alive by hiding them from their enemies. Animals develop coats or shapes that blend in with the background where they live. When they are still, the animals look so much like their surroundings that they seem not to be there at all. A speckled trout blends with the pebbles and water of a stream. A brown sparrow looks like the nest it sits upon.

At the same time, blending with the background, or wearing a disguise makes it easier for hunting animals—the *predators*—to surprise their *prey*. A dark crocodile glides through swamp water looking like an old, drifting log. A white fox, blending with the snow of the Arctic, finds it easier to creep, unseen, close to an unsuspecting hare. Or a sea anemone in a tidepool looks like a harmless flower.

These hide-and-seek tricks are called *camouflage*. It is hard to know what the subject of camouflage might include, as even scientists do not agree. In the broadest sense, camouflage includes several devices—color, shape, pattern, *optical illusions,* warning colors, imitation, and more.

The dictionary says camouflage is behavior, disguise, or skill used to hide or deceive. Most animals do not have to think about camouflage; nature provides this protection *instinctively*. Some animals do not use camouflage as their main protection. Instead, they fly, speed away, jump, or use other actions.

For the most part, camouflage is a trick of the eye. If an animal blends into the background and remains still, it might not be seen. That is, it will be seen by the eye of an observer, but the eye will not distinguish a shape. The eye cannot, therefore, tell the brain that something is there. Seeing—or not seeing—is an important part of camouflage. And what humans see may be different from what other animals see. Just as optical illusions trick people, so may they trick other animals.

Almost half submerged, this Florida crocodile glides on the water like a harmless log.

People have been thinking about animal looks and behavior since ancient times. Charles Darwin was one of the scientists who developed the idea that the colors and patterns of animals helped them survive. He wrote about this idea, and others—*evolution*—in a book called *On the Origin of the Species,* which was published in 1859. The word *species* itself means "kind," and it includes animals that have similar characteristics. For instance, a squirrel belongs to one species, and a chipmunk to another.

Darwin thought that all forms of life were related. Animals, over millions of years, became what they are today because they changed when the world around them changed. As the animals developed, using the earth, water, and sky, their bodies changed to adapt.

Through millions of years animals have been changing, or adapting, like that. Frogs have developed long tongues to snatch flies from the air. Fish have developed gills to breathe underwater. And birds have developed bills to pick insects from hard-to-reach areas.

So many changes have taken place through the millions of years since life on earth began. Since then, a large variety of animals has evolved. Even today, changes are still taking place. Camouflage is one of the reasons why some kinds of animals have survived to the present day. And color, particularly, has played a major role.

2 COLOR DISGUISES

Many animals rely on color to keep them hidden—either for protection or for hunting. Consider the cat family. The leopard is not only fast and powerful, but its coat of yellow with black spots matches wonderfully the dappled forest where it roams. The lion, on the other hand, prefers low shrub and sandy plains. Its uniform coat of tawny brown blends perfectly with the plains background. A tiger with a brownish red-and-black striped coat blends with the grassy plains and reedy swamps it likes.

Of course, not all animals wear the same coat all their lives. Butterflies, for instance, go through different stages. The tiger swallowtail butterfly is a good example. After hatching from an egg, the wormlike young takes on the look of innocent bird droppings. After shedding three more skins, it is green to match the leaves of its home area. In the cocoon stage it merely looks like a brownish broken twig on a branch. Finally, as a butterfly, the swallowtail becomes yellow/orange with black tiger stripes, from which it gets its name.

Another example of color and life changes occurs in birds. Brilliantly colored birds such as parrots and kingfishers build their nests in holes or heavy vegetation. Their eggs and young are hidden already because they are out of sight. Birds that nest on the ground, like terns, larks, and waders, often lay eggs low, in open areas. Many times the eggs blend perfectly with pebbles and large stones. The eggs are hard to see among the natural ground cover. Once hatched, chicks, too, often are colored against their home background.

Although most animals remain the same color during their adult life, some do not. A few cloak themselves in a seasonal costume.

Nature has cleverly helped some animals hide or hunt by matching them with a winter background. Picture a white polar bear on the snow, flattened out, pushing with his hind feet, snaking his way toward a resting seal. Eskimos even insist that a polar bear pushes a chunk of ice in front of himself as he stalks, in order to hide the telltale black of his nose and eyes.

Ground-nesting bird eggs (such as the golden plover's) blend nicely with the ground cover. From any distance at all, the plover bird eggs are nearly impossible to see.

But the polar bear himself is a puzzle. The reason is because he stays white even during summer. Some scientists believe that his white coat is more for saving heat in his body, than for camouflage. The same is true for other animals that stay white all year round—the snowy owl and the American polar hare to name two. So, too, do some Arctic animals, such as the musk ox and the raven, remain dark all year long.

However, several Arctic species do change a dark cover for a white one as winter approaches. The ptarmigan, the mountain hare, and the Arctic fox are some examples. Years ago, scientists thought the change was brought on by cold weather. Now they are not sure. Some people think the number of daylight hours signals the change. During the long daylight of summer, the brown coat is worn. As the daylight lessens toward longer winter nights, the brown sheds to a whiter outfit. Still, not all the scientific questions of seasonal color change have yet been answered.

Above: *a willow ptarmigan in summer coat.*
Below: *a rock ptarmigan in winter coat.*

Every animal's "skin" or outer surface is made up of cells. Some of these surface cells contain a coloring matter called *pigment*. The colors can be black or brown, orange, white, yellow, or red. The pigment cells can combine to make different colors, or different patterns. When these pigment cells stay in one place all the time, they create, for instance, the black and yellow stripes on a tiger. Or the red and brown feathers of the robin. Or the green, black, or brown markings on the cobra. Or even, in humans, the brown freckles on a person's face.

Pigment often changes as an animal grows. The spots on a fawn disappear to a smooth brown when the deer gets older. And the bald eagle does not fully develop the white feathers on its head until it is five years old. Such changes result from shedding hair or feathers of one color, and growing new ones of another color. The changes are slow, and do not change back.

In some animals such as the chameleon, the pigment in special cells can scatter apart or pack together very quickly. When the animal changes its position, different shades or different colors appear on the skin's surface.

Now and then there is another factor involved. To better match fish with the silvery glimmer of

The rocksole flatfish can change color to match its background.

water, and birds with the sparkle of sunlight, some animals have structures within their "skin" or feathers that reflect light rays like tiny mirrors. They give an *iridescent* effect to animal color, like the rainbow shimmer seen in a soap bubble.

What sets off quick color change in the first place?

In many animals, eye vision tells them when they are against a new background. The flounder, a flat fish living in ocean waters, is a particularly clever disguise artist. As an adult it lays flat on the watery ground, both eyes on the same side of its head, looking up. Yet like a submarine with radar, its vision registers enough of the background to take a reading. As the flounder rests in different areas, it can alter its color. Not only can it match colors of sandy, rocky, or muddy sea bottoms, but it can match their patterns as well. Scientific tests show that blindfolded, the flounder cannot change to match its environment. It needs its vision.

While emotions such as fear and anger spark an animal's color for a short period, this is not camouflage. However, several factors beside vision, do bring about camouflage changes. The frog's skin, for instance, will change when it touches different objects, or when it feels different temperature changes. Light also makes a difference in some animals.

But color is not the only important element in camouflage. Nature has several other tricks up its sleeve.

Much of the time animals depend on seeing their prey in order to capture it. If the prey can make itself "unseen" to the hunter, use tricks to make itself "invisible," the prey may often escape. Such tricks are called *optical illusions,* or not seeing what is actually there. That, really, is what camouflage is all about.

Of course, animals do not go to school to learn these tricks. After all, they cannot reason like humans. Instead, they practice these camouflage actions without thinking, instinctively. But scientists have studied camouflage, and they recognize many of the tricks animals play to fool the seeing eye of their predators.

Since the sun brings light each day, everything on earth that is solid, or that you cannot see through, casts a shadow. In bright sunlight, the shadow is dark. When the sky is overcast, it is hardly noticeable. But animals are aware of shadows. For where there is a shadow, there is something solid as well. That is one thing hunting animals look for. Shadows show an outline which gives further clues. You know this is true if you have ever cut a silhouette, or copy, of a famous person's face from black paper. You can still recognize the person. Animals can often recognize a prey by such an outline, too. So if a prey can make its shadow or silhouette disappear, it may escape its hunter.

To see how shadows work, try this experiment on a piece of paper. Draw a circle. It is flat, there is no shadow. With dots, put in a shadow like this.

Now you will have what looks like a solid object—a ball. The shadow gives the circle depth.

Some animals instinctively know that by doing away with their own shadows they will be harder to see. The closer an animal hugs the ground, or the object it sits upon, the smaller will be its shadow.

A racing crab, for instance, is well camouflaged in color against its background at the beach. Yet it is often moving about. The crab's shadow, of course, follows along and betrays its whereabouts. The crab seems to know this. When it senses danger, it scuttles to a hollow in the sand to hide the telltale shadow.

The best way to eliminate a shadow is to be transparent. A South American butterfly throws practically no shadow because light shines right through its clear wings. And there is a species of squid that lives in the water and gives off its own light, much like a firefly. The squid shines its light on its shadow, thus doing away with it. This squid can even darken or lighten its light depending on the strength of the sun entering the water.

Many scientists believe that another way nature helps trick the eye with shadow is through *counter-shading*. Most mammals, a great number of birds, and practically all species of fish that are not bottom dwellers are dark on the top and light under-

This mew sea gull displays countershading with its gray back and white underparts.

neath. Lions, mice, otters, owls, and herring are some examples of such animals. Closer to home, you can even see such shading on your own dog or cat.

In nature, as was mentioned earlier, light mainly

comes from one direction—above, from the sun. Light shining down on a dark object tends to lighten it. At the same time, light from above shadows an object underneath, or away from the source. Lighting a darker color from above, and darkening a lighter color underneath, tends to mix the two colors. From a distance, these two colors seem to blend

Alaskan murres are strongly countershaded like many fish.

into one tone. In black and white colors such as in ocean fish, a blending shade of gray is produced. This almost makes the fish ghostlike in color. By doing away with shadows and mixing the colors, the object seems to flatten and merge with the background.

To see how this process works, look at this chart:

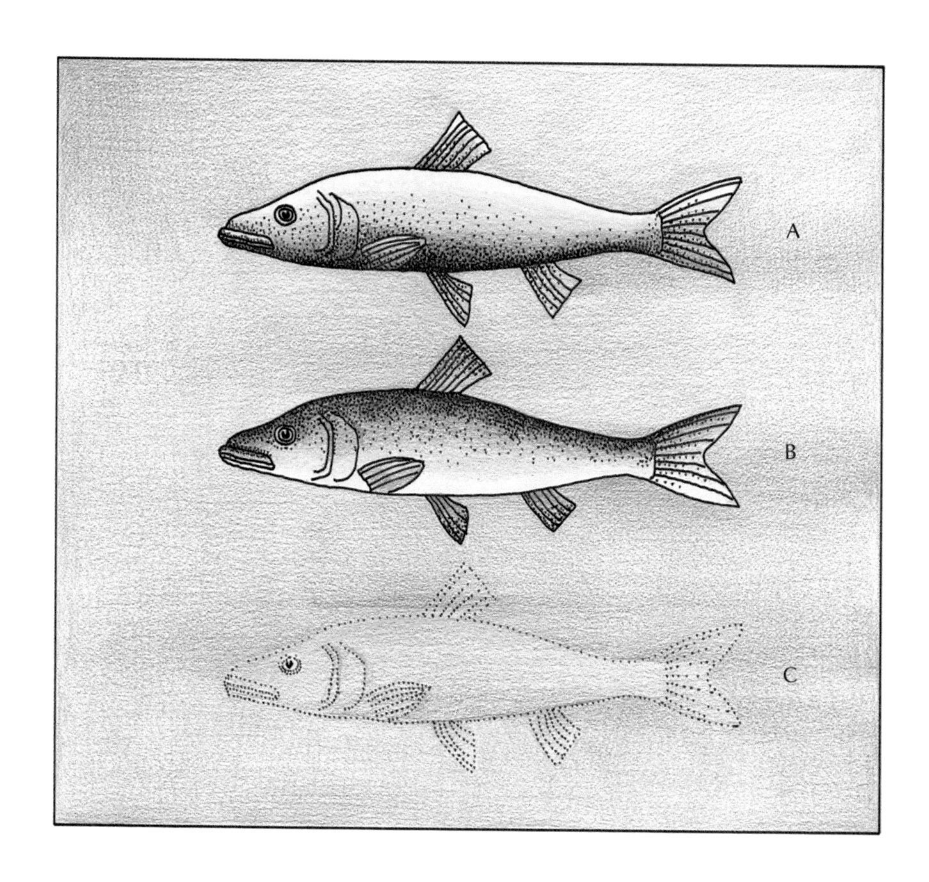

In (A) you can see how the sun lightens the top of a body and throws shadow underneath. (B) is an example of the coloring of a countershaded fish. The combined effect of top lighting and countershading is shown in (C).

Countershading is particularly useful for animals that live in water. An eagle hunting salmon from above will find it harder to spot the fish because of the salmon's dark upper surface against the water. The light from above shines down on the dark fish and lightens it, helping it to blend with the water. On the other hand, a fish looking up from underneath at another fish has an equally difficult time. All the predator fish sees is the light shadowed undersurface of its victim against the brighter light of surface water. Of course, this countershading trick does not always work, but it helps.

And then, there are a few animals that are different. For instance the Nile catfish of Egypt, which feeds on plants that grow on the surface of the water. Interestingly, it swims upside down in order to do this. Nature, however, has done its part to protect this rebel. This fish is shaded light on the top and dark on the bottom—just the opposite of other fish. The catfish can swim upside down and still carry its protective shading.

As much as animals try to hide their shadows, sometimes shadowlike markings on animals help to camouflage them even better. Often brown or black markings on birds, moths, snakes, lizards, and other animals suggest fake shadows. These fake shadows help the animals to blend into their patchwork backgrounds. In this way, shadows, or make-believe shadows, help an animal hide.

Underwing moths are especially good examples. Usually they have whitish, gray, or brown forewings, which are patterned with lacy light and dark markings. When at rest on the bark of a tree, the moth adjusts itself to line markings with the markings on the bark.

The camouflage tricks of nature are not over yet. It has another optical illusion to present. This one "breaks" an animal apart.

The underwing moth matches its marks with that of tree bark.

4 THE DISAPPEARING ACT

Some animals are the same color as their background: yellow snakes slithering on sandy soil, black bats flying at night, green caterpillars crawling on a leaf. Frequently mammals, insects, and birds are the same color as the ground or environment where they live. Take away their shadows, and they are hard to see.

But not all animals are one color. In fact, many animals and insects are many colors and patterns. Look closely at frogs, for instance, many of which have bright stripes on a darker background. Some animals, like the jaguar, have dark spots, or like the giraffe, have uneven patches. Snakes display a variety of patterns.

Tricking the eye also helps in camouflage when an animal's markings blend with the background. If

The bold patterns worn by zebras and giraffes are not as easily seen when the animals mix with their brushy background.

an animal is marked in such a way that its lines or colors follow into its surroundings, the viewer's eye will follow, too. If the animal has lots of lines or stripes or spots that match the background it is seen against, its outline will be harder to see.

Such an illusion works like this. When an animal is covered with irregular patches of different colors and tones, these patches tend to catch the eye of the viewer. When they do, they draw the viewer's attention away from the actual shape of the animal. The eye tends to read the bold patches as separate objects, as if they were broken apart. This principle of camouflage is called *disruptive coloration.*

Perhaps one of the best examples of disruptive coloration in nature is the shockingly bold stripes of the zebra. You might think zebras would be easily seen in nature. In sunlight, in open country, they are. But from a distance, near cover, it is hard to tell the shape of a zebra. And at night the bold stripes blend so well with the starlight shadows that zebras are nearly impossible to locate.

Bold patterns, you might think, would be out of place on baby birds who have few defenses and just want to hide. In many cases this is true. But not always. It depends on the background where the chick is living its first weeks.

As an example, take ringed plover chicks. They have black and white bands circling their downy heads. Against a background of pebbles on a riverbank, they are extremely difficult to see.

You can see how disruptive coloration works against different backgrounds in the following chart. Notice how the outlines of the objects often lose their shape.

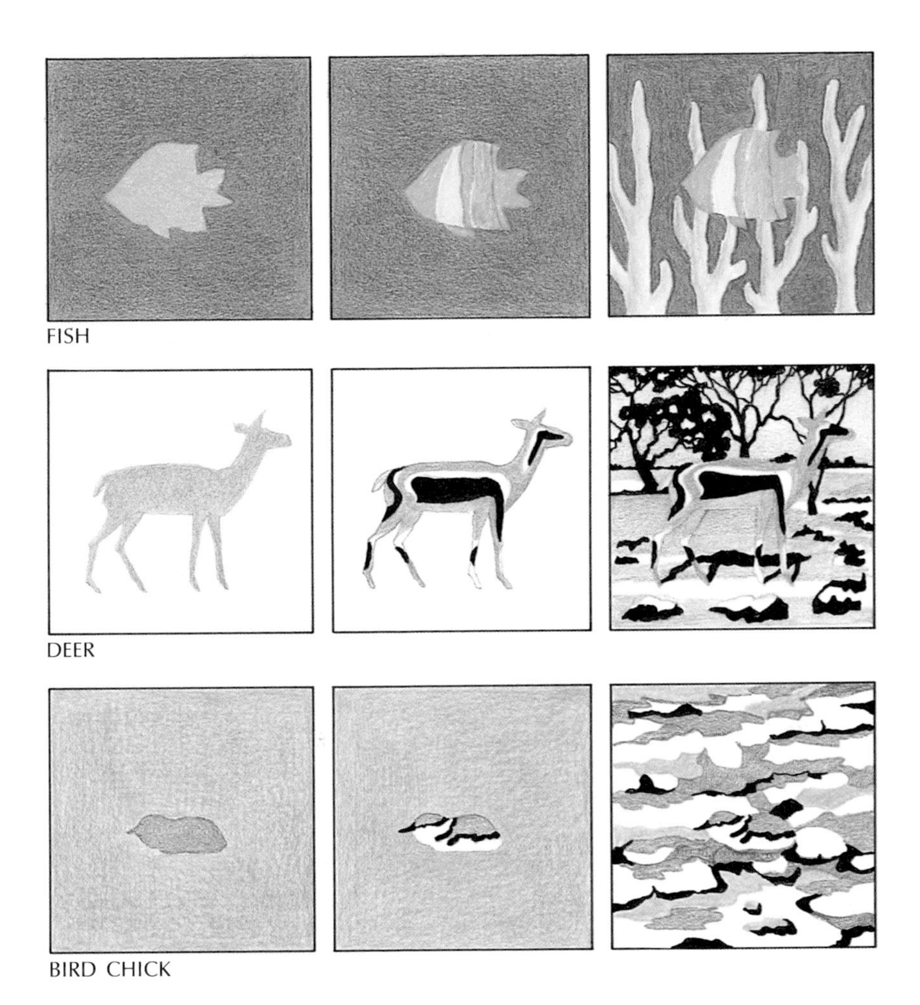

Disruptive coloration plays another part in nature, especially when it concerns the tiny feature of the eye.

Eyes are hard to hide. Few things in nature are as round as eyes, so they often are recognized when the rest of an animal remains concealed. Mammals, birds, fish, reptiles, and numerous insects may be different in shape, but, for the most part, the eye remains round on all of them. Not only that, but the eye is normally on a critical part of the body where other important features are—the nose, the mouth. If a hunter can do harm where the eye is, the prey will often have other important functions damaged. That's why predators often attack the eye first.

Nature has its own tricks, though. Dark stripes, masks, and other markings on the animal's face help hide the eyes from view. For example, a variety of fish and snakes have black stripes on their bodies. Sometimes the stripes include the dark eyes, making them very hard to see.

But nature has gone even further to protect its creatures. It has fashioned "fake" eyes, or eyespots, on less important parts of an animal's body.

Depending on their size, eyespots may serve several purposes. If they are large and suddenly

With the black slash over the real eye, and the large fake eye near the tail, this butterfly fish almost looks like it is swimming backwards.

shown, they may surprise a predator and scare him away. Certain caterpillars have two large make-believe eyes at the top of their heads, which make them look like fearful snakes.

But the eyespots usually located away from important parts of the body are believed to be used for camouflage. Scientists have studied some butterflies with eyespots on their outer wings. When exposed to birds, the butterfly sometimes lost a wing section with the fake eye; there was a triangular beak mark where the bird took a bite. The eyespots were the target instead of the real eye. But the actual body of the butterfly was not harmed at all.

After using all kinds of tricks to hide, it is strange to find some animals that *want* to be seen. Now let's look at the splashy animals—the advertisers.

5 THE FLASHY ADVERTISERS

When you are watching television, notice the advertisements. Often you will see a lot of color and movement, or will hear singing or talking which is meant to catch your attention.

Nature has advertisers, too; animals that are usually bright red or yellow or orange. They show up against a nonmatching background like video games on a TV screen.

The monarch butterfly is one advertiser. It is orange and black and flies in the sunlight, rests against green leaves or brown bark, and does not seem to care in the least if it is seen by other animals. It's almost as if the butterfly is saying, "I'll go where I please, and I don't care how showy I look. Go ahead and take a bite. But you'll be sorry!" The reason

Because it is foul-tasting, the monarch butterfly does not care whether it is seen or not.

the monarch might send this message is because any animal who bites it is due for a mouth full of nasty-tasting stuff. It doesn't take many mistakes for birds or frogs to pass up a monarch meal.

You might wonder why a book on camouflage would deal with such bright advertisers as the monarch butterfly. But bright coloring, in a way, is a kind of camouflage. It protects an animal and helps it survive by warning away its enemies. Also, there are some species that look the same as the bright creatures, yet taste delicious. Such look-alikes are called *mimics,* and are protected because they look like the advertiser. Many a hunter will not touch them. More about mimics later.

Bright colors do not always mean danger to another animal. Many males display themselves through color or behavior while defending their territory. A number of butterflies and birds are brilliantly colored for the purpose of mating and raising families. Usually it is the male of the species who is flashy, while the female keeps her dull coat to blend with her hidden home. Bright fishes along tropical coasts are easily seen by people. But such creatures may be better camouflaged from their enemies who see them with a fishy eye among the shimmering shallows.

Warning colors say, "Watch out!," but some animals do not listen. If a hunter does bite, he is likely to get the second line of defense—an irritating or poisonous sting, or a nasty odor. That white stripe on a skunk's back warns a hunter to stay away, but the skunk's odor has the final say.

While frogs are usually green and brown, there is a brightly barred red and black frog in South

Right: *the poison arrow frog pictured here is bright red and blue to warn away its predators.* Below: *with spines close to the body, the porcupine fish does not appear alarming. Inflated, however, it is quite another matter.*

America. Scientists found that the skin fluid on its back was the source for the poison Amazon Indians used on the tips of their arrows.

Another feature of bright advertisers is that they may move slowly. For example, when the pink and black Gila monster lizard is in danger, it just stops its leisurely crawl and looks at the enemy. The Gila monster knows it can rely on its grooved teeth and not worry about running away. After all, with deadly poison to bite with, who needs to hurry?

Most animal advertisers move around by day. However, there are a few advertisers who are active at night. They either carry black and white warning patterns or use other warning signals to protect themselves. Smells and sounds are very effective at night. For instance, a porcupine not only displays his black and white quills, but as he waddles along, his chorus of needles sings a warning song. And most predators listen very keenly.

The spiny balloonfish has a special defense. When he is fearful, he puffs himself up like a ball, and the spines on his body bristle out. Only a hunter fish with an appetite for pincushions would dare bother him!

With all the care flashy animals are shown, it is not surprising that some harmless creatures benefit from this protection, too.

around, and probably flies, too. But when an insect is attracted to your red shirt and starts buzzing close to your head, you quickly duck and swing your arms around to wave it away. You do not calmly stop and try to examine the insect to see if it has one set of wings or two. No, you act first. Some hoverfly hunters may have the same reaction. They will avoid the hoverfly, thinking it may be a bee. And they won't get too close either.

Not all models are colorful. Ants, for instance, vary from reddish brown to black. Yet because some species bite, sting, or contain bad-tasting fluids, a great number of predators avoid them. Because of this, a lot of insects and spiders mimic ants—right down to their busy bodies and dull colors. Copycat spiders even position their front legs above their heads so they look like the antennae of ants.

Just because a model is brightly colored or tastes bad does not mean it's safe. Predators learn to leave certain insects alone, but some disagreeable models are eaten before the lesson is learned.

The hover fly (top) looks so much like a stinging honey bee, it gains protection because of its color and pattern.

Examples of mimics are found in the ocean, too. The model, the sea swallow fish for example, helpfully eats parasites from larger fish. But the mimic, an aggressive blenny fish looking much like the sea swallow, uses its bright blue, black, and white coloring to feed upon unsuspecting fish. When a customer fish comes for cleaning and the copycat blenny is playing the role of cleaner, the large customer fish will have a bite taken *out* of it rather than *off* of it.

Not all mimics use animals for models. The orchid mantis of southern Asia resembles a flower. Before it is fully grown, this insect is pink in color, with broad hind legs that resemble petals on a flower. When hungry, the mantis crouches among real flower petals just waiting for its meal to come by—a bee or a butterfly. The mantis is so attractive to insects that it lures them even when it is sitting on green leaves all by itself.

The aggressive blenny fish (top) takes advantage of its look-alike coat, and goes after the cleaner wrasse's customers.

LET THE SHOW BEGIN

Now to the stars of the show—the impostors, those who make themselves "invisible" in unusual ways.

There is a fish in the lower Amazon Valley of South America that looks like a dead leaf. In fact, it is known as the leaf fish. It glides on currents, head down, or lies on the bottom of a stream. Its body is flat and leaf-shaped, with a skin pattern similar to mildew on a dying leaf. In the water its body is held stiffly or slightly curled. But should a smaller fish come within reach, the "leaf" darts into action, gulping its prey before it knows what happened.

Other animals, such as toads, resemble leaves, too. But the most effective leaf imitators are among the insects. They have a head start over other animals because their wings are thin and veined like leaves. Many have sticklike legs similar to twigs on a bush. Treehoppers, for instance, resemble pointed thorns on a branch. Butterflies and moths get top billing when playing the part of a leaf, while some caterpillars star as twig-imitators.

A number of caterpillars are colored like their favorite plant—reddish purple, green, or brown. They often grow bumps on their bodies that resemble buds of a twig. But their behavior is most remarkable. Instead of crawling along a plant all the

This Malaysian insect looks like the leaf it sits upon.

time, they grasp a twig with their hind legs and stiffen out away from the branch, looking much like a growing twig. Some species even bend a little to make the disguise more convincing. When birds are searching for food, they cannot try every twig to see if it is real, and so the caterpillar often wins the hour.

Marine waters, too, abound with animals disguised as leaves or seaweed or ocean grass. Pipefish, long and narrow, look like the sea grasses they live among. They even swim vertically, head up, to complete the picture.

Some animals use props. Crabs and insects collect objects from their surroundings and pack them around on their bodies. The animals are very careful about how such material is arranged for the best disguise. The spider is an especially clever creature. It not only often sports a color camouflage, but some species mimic ants, and receive protection that way. Its web is used both as a trap to catch food and as a backdrop in which to disappear. The spider has many tricks.

Using their web as a prop, orb-weaver spiders sit in the center or hub. They then stretch their first two pairs of legs out along the web threads in front, and the back two pair along the web behind. When the spider is still, it looks very much like the web itself.

Another kind of spider gathers a dead curled leaf and places it in the center of its web. The spider then hides in the leaf.

Some South American spiders carry a dead ant on their backs. The scene looks like an ant carrying an ant. The spider not only looks the part, but it runs around in a jerky fashion like an ant.

A caterpillar of Borneo attaches flower buds to its back, and changes the buds to fresh ones as they dry up. A weevil in New Guinea actually grows plants on its back. It looks like a walking garden.

A caddisfly larvae lives underwater. It spins a "mobile home" of silk, and glues stones and sand to it. The home makes a camouflaged place to live.

Certain masked crabs pick up sponges from around them on the ocean floor. They cut the sponges to correct sizes with their pincers, chew them for a few seconds, and then rub them onto their bodies. Before long, this "mask" covers the outer shell of the crab.

Perhaps the star in this production is the carrier shell which lives in the sea. One species gathers empty shells from the shallow water and very precisely places them upside down on its own shell. The carrier uses a substance from its body to "glue" the piece in place. When a big piece has been attached, the carrier will stay still for as long as ten hours to be sure the piece is firm. Then it will continue to add shells. All finished, it looks like a pile of rubble on the sea floor.

When all else fails, a number of prey try to discourage hunters by making believe they are already dead. There are some predators like cats and mantid insects that attack living animals only. The trouble is, too few hunters are discouraged by a "dead" animal. They just go ahead and eat it anyway.

Rolling over and playing dead like this hognose snake does not always fool predators.

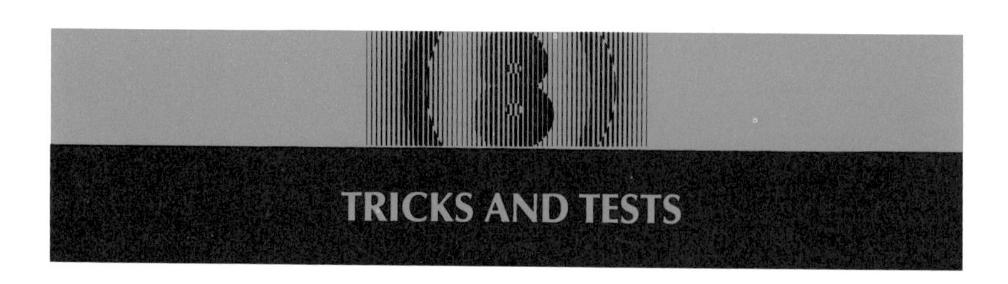

TRICKS AND TESTS

Since tricks of the eye play such an important role in camouflage, test yourself to see if your sight can be fooled.

Look at these two lines. Which line is longer?

Now take a ruler and measure each line.

What did you find? Did the outer lines fool you

into thinking one line was shorter than the other? Did you guess the right answer?

Scientists, too, can only make some very good guesses when it comes to the subject of camouflage. There are several reasons for this. In the first place, humans have a complicated camera-type eye, which is different from the compound eye of insects. It is difficult for scientists to know what an insect is actually seeing.

Second, humans can see in color, as do birds, reptiles, and some other animals. Some animals, however, do not see in color, especially animals that hunt by night. Those animals depend on other signals such as smell or heat to find their prey. They rely on the brightness of their preys' coats to find them. In addition, some animals may have a wider or different vision range—maybe including infrared or ultraviolet, which humans cannot see. Only scientific tests can answer the many questions about camouflage. Some of those tests are very difficult to perform.

In spite of the problems, scientific experiments have been carried out. The purpose of one test was to see if tiny fish—minnows—were fooled by an optical illusion. If so, the tests might tell scientists a little more about what fish are actually seeing, and if they are fooled by camouflage tricks.

The experiment was performed by Dr. Konrad Herter. First, he set up two rectangular black blocks. One block was larger than the other. Then he taught the minnows to feed near the larger block.

After the fish were trained, Dr. Herter replaced the blocks with an optical illusion. It looked like this:

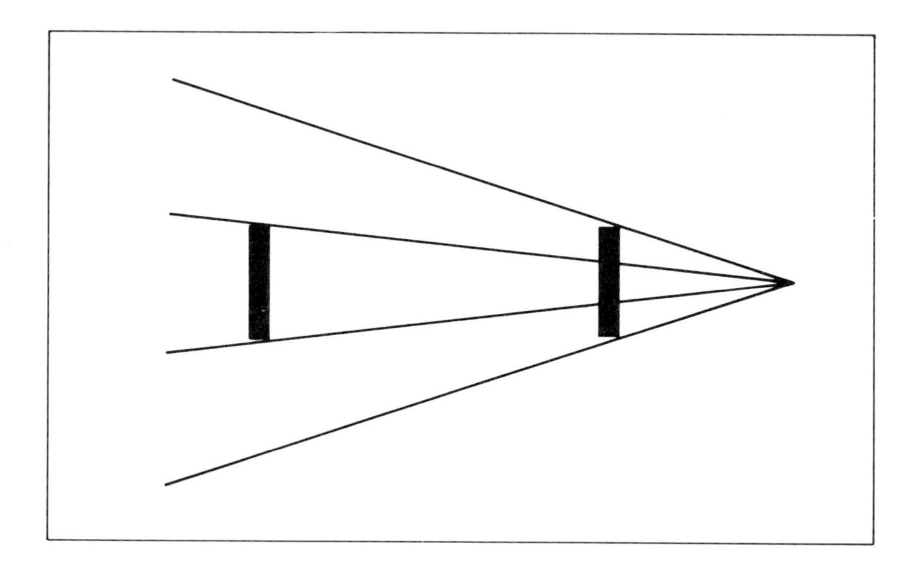

Both of the blocks were the same size, but one *looked* larger because of the lines in the background. During the experiment, the minnows chose to feed near the larger-looking block more often than not. Other tests with optical illusions gave similar results. From these tests you might think that humans and fish are both fooled by optical illusions, and by camouflage tricks as well.

Nature has many clever ways of protecting animals. Camouflage is only one way.

You don't need to travel to distant lands to see such wonders. Animals are everywhere—in the woods, on the beach, in a park—even in your own backyard. Just look around.

GLOSSARY

Camouflage—a disguise or a behavior that is used by an animal to hide itself or deceive an enemy.

Countershading—is the way many animals are colored darker above and paler below to counteract shadows caused by overhead daylight.

Disruptive coloration—a boldly patterned coloring that breaks up the shape of an animal, making it harder to recognize.

Evolution—the development and changes in animal species occurring over millions of years. Evolution changes the characteristics of the species and can often lead to a new species.

Instinct—the impulse of an animal to respond automatically to something.

Iridescence—a play of colors that produces rainbow effects.

Mimic—an animal that looks or acts like another animal, or plant, or rock.

Mimicry—an animal uses mimicry to protect itself against enemies. One animal (the mimic) looks like another animal (the model) or plant, or rock, etc.

Model—an animal that is imitated by a mimic animal. The model usually has disagreeable characteristics (smell, taste, etc.) for predators. The look-alike mimic benefits from the resemblance.

Optical illusion—an image seen by the eye that is not what it appears to be.

Pigment—a colored substance in the cells and tissues of plants and animals.

Predator—an animal that kills and eats other animals.

Prey—an animal that is hunted and killed by another animal for food.

Species—a group of animals with similar features that can breed to produce young.

FOR FURTHER READING

Clarkson, Jan Nagel. *Tricks Animals Play.* Washington, D.C.: National Geographic Society, 1975.

Cole, Joann and Wexler, Jerome. *Find the Hidden Insect.* New York: William Morrow and Company, 1979.

Pringle, Laurence. *Chains, Webs, and Pyramids.* New York: Thomas Y. Crowell Co., 1975.

Ross, Edward. *Camouflage in Nature.* Chicago: Children's Press, 1961.

Shuttlesworth, Dorothy. *Animal Camouflage.* New York: Natural History Press, 1966.

Simon, Hilda. *Feathers Plain and Fancy.* New York: Viking Press, 1969.

———. *Insect Masquerades.* New York: Viking Press, 1969.

Simon, Seymour. *The Optical Illusion Book*. New York: William Morrow and Company, 1984.

Young readers' magazines such as *Ranger Rick, National Geographic World,* and *Owl* occasionally carry features about animal camouflage.

INDEX

Page numbers in *italics* refer to captions

3 9004 02355003 8